The Human Rights of Hosts on *Westworld* and the Wave Chicane

Paul Levinson

CONTENTS

INTRODUCTION

I love paradoxes, and *Westworld* is full of them. When I reviewed the first season (Levinson, 2016), I noticed some paradoxes that typify and animate the entire series.

One is well-known in philosophy and communication theory, and most associated with Paul Watzlawick (1976): the "be spontaneous paradox". If someone asks or commands you to be spontaneous, and you comply, you are not being spontaneous. But if you ignore or disobey the request, and deliberately act in a non-spontaneous way, you are also are not being spontaneous. Hence, the paradox. In *Westworld*, this happens any time any of the hosts display an iota of originality. They of course are not being original, if they have been programmed to act in that way. On the other hand, maybe the originality was not the result of the programming, but a malfunction. Dolores's killing spree in the second season brings home this issue, and occasions a third possibility: maybe she is evolving into something that goes profoundly beyond her programming. Though perhaps such an evolution can be considered a kind of malfunction.

Another kind of paradox is one which I came up with on my own, back when I first watched Data on *Star Trek: The Next Generation*. The problem with all the androids and human-like robots we usually see on television and in the movies is that they are portrayed by human actors. We cannot help, as human viewers, to viscerally respond and relate to those robots as if they were human, which begs the question in the narrative of how human or not these artificial beings may be. In *Westworld*, Dolores and Maeve are played by beautiful, sensitive actors. When we strive to assess the humanity in these host androids, we cannot help but be influenced by the human beings portraying these characters, and the humanity or humanness they inevitably project.

But my favorite paradox in *Westworld* is one which I also came up with, some years ago, as I pondered the robot stories of Isaac Asimov. As we develop androids which are ever more intelligent to do our bidding, at what point have we created sentient beings to be our slaves, which would be as immoral for these human creations as it is for human beings. Every host in *Westworld* is at one level or another on that morally paradoxical stage. And, indeed, the more advanced the host, the more they pitch we their human creators into that quandary. This pertains most to Dolores and Maeve, who begin to discover their need for independence in Season One, and act upon that in Season Two, and the Man in Black, who we discover at the end of Season Two is likely a host. It also pertains, in a very different way, to Bernard.

THE HUMAN RIGHTS OF ROBOTS

I first addressed this dilemma in 1998, in a brief essay published in *Shift* magazine, the short-lived Canadian *Wired*. I entitled the essay "The Civil Rights of Robots," because I thought that was catchy. But "civil" does not go far enough. What we're really talking about regarding sentient beings we create are their human rights. I suppose human rights are implied in civil rights, but we might as well be up front about it.

The key ingredient that qualifies an AI (or robot or android) for human rights is sentience – or more precisely, the degree of sentience. The anti-collision and like features in my 2017 Prius are a form of AI, but no one would say my car is sentient. So if, for example, my car refused to let me drive it to a gas station, because it didn't want to go out in the blistering heat, no one would say it was a sentient being exercising its human right not to be ordered to do an unpleasant thing. Everyone including my Toyota Service Center would agree the car was malfunctioning, and required some kind of repair.

Of course, my car does not only look nothing like Dolores or the Man in Black, it doesn't look human at all. This returns us to the paradox of a talented human actors portraying androids on television. But if I owned, say, a vacuum cleaner with AI, which looked like a human being, we would likely still be able to easily conclude that it was not entitled to human rights, and if it didn't perform the way we expected and instructed, we would bring it in for repair, or, if unrepairable, toss it in the scrap heap, without a trace of a guilty conscience.

The very word "owned" shines a spotlight on what we are dealing with here. I own my car and my vacuum cleaner, at this point whatever the AI that may be installed. Manifestly that's not the case with Dolores or Maeve. *Westworld* may legally own them. But what we see of their behavior makes it patently clear that such ownership is as immoral as a human being owning another human being.

But what is it in the behavior of some of the hosts on *Westworld* that would lead us to this conclusion? And should it make a difference if such behavior is deliberately programmed, or the result of an unintended evolution of the original programming?

THE PRECONDITIONS OF SENTIENCE

Regarding sentience arising from biological evolution – which is how ours arose, as far as I know (meaning, I'm assuming extra-terrestrial or divine intervention was not the source) – I'm willing to go out on a limb and say that we humans are thus far the only organisms that have it. Though the genomes or chimps and great apes are more than 98-percent the same as ours, they have not yet produced a Jonathan Nolan or a Lisa Nolan, who co-created *Westworld.* The same for dolphins. I have no idea what they're thinking or may be saying, but if remaking the world via the invention and application of technology is a signal characteristic of sentient intelligence – which I would argue it is (see Levinson, 1988) – then dolphins fare no better or even a little worse than chimps.

But though human intelligence is apparently unique in our biological world, it is indisputably a part of, and arose from and in, that biological ecosystem on our planet. This is why I've always thought that Maurice Merleau-Ponty (1962/1981; see also Edgar, 2016) made a crucial point when wrote of the "metaphysics of flesh," or the inextricable necessity or presence of life for sentience and thought.

What I've always taken this to mean is that sentience cannot exist in a biological vacuum, or in the absence of a package or vehicle which is alive, and has the properties of life. This would rule out one of science fiction's favorite gambits, a sentient intelligence in a huge mainframe computer. To coin a phrase (actually,

I coined it in Levinson, 1988), that would be a case of putting Descartes before the horse, or saying mentality existed before life. But a robot or an android which moved through the world in the same way as human beings and other living organisms would be a more likely locus of sentience.

Note that I don't think that we need to insist that DNA be the source of the body that is sentient. A digital code embedded in silicon or whatever non-flesh format might well suffice as the basis of an entity we would agree would qualify as living, and the basis of sentience. To insist on DNA as the only format for such life and intelligence would amount to a protein chauvinism. Thus the positronic brains in Asimov's robots (e.g., Asimov, 1957) and the bi-cameral circuitry in *Westworld*'s hosts would both qualify.

The bicamerality of the hosts' brains – the two halves of the host brain talking to each other, and therein attaining sentience – was first mentioned by Dr. Ford as early as episode 1.3 of *Westworld*. The concept of bicameral minds was first developed by Julian Jaynes in his *The Origin of Consciousness in the Breakdown of the Bicameral Mind*, which achieved considerable attention when it was published in 1976, and has since become something of a cult classic. The book argues that in the distant past, human beings literally had two minds, located in each hemisphere of the brain, which communicated or talked to the each other. The remnants of that in our modern world are accounts of prophets in the past who heard voices, and schizophrenics who continue to do so. The rise of literacy which focused the mind resulted in the unified single consciousness which we have today.

I met Jaynes at more than one academic conference, and worked with Marshall McLuhan, who did extensive work on the rise and impact of literacy (e.g., 1962). I thought (and still believe) that a crucial weakness in Jaynes' formulation was no evidence of lack of consciousness in people who are illiterate in the modern world, including pre-literate children. I once asked McLuhan what he thought of Jaynes's theory, and he replied, "science fiction". Like so many of McLuhan's observations which in effect were almost prescient – such as the global village (1962) nearly half a century before the World Wide Web (in 1962, there were national no international audiences via television, and they were not able to interact on an easy global basis; see Levinson, 1999, for more) – his take on the bicameral mind foresaw the way the concept would be used in the science fiction that is *Westworld*.

Of course, in *Westworld*, consciousness or sentience is created via the internal interaction of the bicameral mentalities, in contrast to consciousness emerging from the breakdown or destruction of the bicameral mind due to literacy, as per Jaynes. But, close enough, and in the bicameral precondition, *Westworld* has provided a path via which sentience could emerge. That it indeed did emerge in some hosts is evidenced by the access we are given to their internal thoughts, and the actions they take against their programmers.

This second criterion is given bright and violent exposition in the second season via the rebellious rampage of Dolores. A cheap, unsatisfactory take on this would be that something went very wrong with her programming, and she has turned into a homicidal maniac. But if she is sentient, she becomes a kind of Spartacus, unleashing death not entirely undeserving upon her captors.

THE MORALITY OF SLAVE REBELLION

The "thou shalt no kill" commandment has already received ample qualification in our real world. It's ok, even ethical, to kill an enemy in wartime, or in any life-and-death defense of you or your family. Is it ok for a slave to kill her or his master?

We might say, in our real world, that a better, more moral option is to escape without violence, if possible. But that option is not clearly available, if available at all, to the hosts in *Westworld.*

The hosts are also not just slaves. In the case of Dolores and Maeve, they have been repeatedly abused, raped, and murdered. In the case of Maeve, she also been deprived of her child, or perhaps her programmed memory of a child, which amounts to the same kind of psychological cruelty. Of course, if neither host is sentient, then these abuses are for the most part mitigated. We regularly deprive our pets of their offspring. And although abuse of animals may be immoral or illegal, such abuse is in a league completely different from abuse of a human.

I acknowledge that animal rights activists might disagree, i.e., see no difference in principle between abuse of animals and abuse of humans. But agreement with this argument strengthens not weakens the claim that the hosts on *Westworld* have been victims of intolerable abuse, because an equality between animals and humans would make *Westworld* host abuse outrightly unacceptable, even if they in some way were less than humanly sentient.

To return to our traditional standards of ethics regarding acceptable actions of slaves abused by their masters, the story of Moses and the Jewish exodos from Egypt is instructive. At the Passover seder, the Ten Plagues that God brought upon the Egyptians at Moses's request, when the Pharaoh refused to release the Hebrews from bondage, are recounted. They conclude with the slaughter of the Egyptian first-born sons – a harsh even heinous action which finally gets the Pharaoh to relent.

An atheistic history account of the same story might well substitute Jewish freedom fighters for God, or certainly enactors of the will of God. Are Dolores's actions any different? If not, and if we agree with the tenets and lessons of Judao-Christian history, then Dolores's killing spree is neither immortal nor even excessive, but a warranted if extreme response to the humans who enslaved her, and continued to do so.

Of course, one difference between Dolores's actions and the killing of Egyptian first-born males as recorded in the Passover *Haggadah* is that Dolores killed not only the people who enslaved and continued to enslave her, but created her. This makes the programmers on *Westworld* even more culpable that the Egyptian slave master masters, since, unlike the Egyptians, the programmers deliberately created the hosts to be hosts or slaves.

Whether the programmers or those who hired and instructed them knew they were creating hosts to be as sentient as Dolores and Maude is yet another question. But I think the answer does not matter: whether by intent or accident, the existence of sentient hosts, and thus how to treat them, is the responsibility of those who created and programmed the attractions at *Westworld*, the attractions being the hosts.

But this does raise the important question of, given that the hosts have achieved sentience, could new hosts be programmed to be less sentient, or not sentient at all, and yet perform the desired services?

ARE THERE ANY MORAL, NONPARADOXICAL PATHS FOR THE PROGRAMMERS?

Deciding how much sentience warrants being accorded the same decency as humans – that is, not being subject to all manner of abuse – assumes that there is such a thing as fractional sentience. One might well argue that anything less than human sentience, any fraction of it, however large, is not sentience at all. Given what I said above about no chimps or dolphins producing anything like *Westworld* or Asimov's robot stories, it should come as no surprise that I'm inclined to agree with this line of thought. In other words, however intelligent dolphins and chimps might be – and they certainly are, in comparison to other animals, not to mention plants – this does not qualify them as sentient.

But this in turn returns us to the question of pets. Dogs are apparently not as intelligent as chimps, and we humans have made pets of them, or at least have had a relationship with them in which they do tasks for us like herding, for at least 15,000 years. Animal rights activism has ranged from urging that dogs and other pets be treated humanely (this is sometimes referred to as animal welfare) to the absolutist position of no pets at all (the Wikipedia article "Animal Rights" is comprehensive, and has a plethora of citations).

If we accept the position of no pets at all – that owning any living creature is immoral, and commanding it to do our bidding even more so – then this would rule out any hosts or androids at all at *Westworld*, unless the fact that the hosts had circuitry rather than DNA made them nonliving and thus not subject to the pet prohibition. As I indicated earlier, I would reject this position as protein chauvinism. But if we do think hosts are entitled to the same consideration that animal rights activists give to dogs, we are confronted again with the question of how much sentience or even just plain intelligence (such as that evidenced by dogs) would be required by animal activists to extend such protection to androids?

The extreme position of all animals being entitled to freedom, or not being anyone's property, wouldn't work for all of our devices that are now embedded with any intelligence. Driverless cars might create moral dilemmas for humans who use them (see Walton, 2019, for a riveting novel that addresses this issue), but such dilemmas – such as should a driverless car avoid serving into another car to protect its passengers but in so doing hit another car and endanger its passengers – concern the welfare of sentient humans not their cars. (I'm tempted to say that an extreme animal activist might argue that driverless cars must also avoid hitting any squirrels.)

But I'm sure most people would agree that a squirrel's life is worth sacrificing to save a human by a driverless car, or indeed by any machine, possessed of AI or not. And this puts on the table not only the caliber of the "slave" – is it a dog, a squirrel, or an insect, is it a driverless car or a *Westworld* host like Dolores – but the purpose for which the slave animal or slave host is being employed. I would certainly say that a dog which lost its life following its master's command to save a child in a fire was not being mistreated. If this makes me a human chauvinist, then so be it.

But the hosts on *Westworld* are at best ministering to the mental well-being of guests, not saving their lives, other than in fictitious role-playing. If we agree that at least some of the hosts on *Westworld* are sentient – certainly Dolores, Maeve, Bernard, and likely the Man in Black – then they would constitute a classic, ipso facto use of slaves for non-life-and-death purposes. Indeed, the cases of Dolores and Maeve, for at best frivolous and at worst blatantly immoral purposes such as non-voluntary prostitution, would be indisputable instances of this.

But what of the Man in Black? If he is a host, he is clearly not in the same league as Dolores and Maeve, who from the moment we met them were straight-up slaves. In contrast, the Man in Black was and is so different, that it wasn't until the last scene in the last episode of the second season that we even learned that he very likely is a host.

But a very special kind of host.

HOW THE MAN IN BLACK FITS INTO THE QUANDARY OF HOSTS AS SLAVES

The gist of the Man in Black is that he likely was a guest who became a host. Or, based on the scant evidence of the final scene in the Season Two finale, perhaps he is a guest who will become a host, at some time in the future. Either way, as far as we know, his becoming a host was more of a voluntary act than for any of the other hosts, including Dolores and Maeve, and, as far we can tell, Bernard.

So the first question that the Man in Black raises for our ethical journey is: if someone or some entity volunteers to be a slave, does that justify anything his or her owner might require the slave to do? Perhaps. But this situation is further complicated on *Westworld* by the fact that the Man in Black does not act like the other hosts, including Dolores and Maeve. He is not used for sex, is not regularly killed and brought back to life, etc. (Bernard shares at least some of these lack of "normal" host characteristics.)

Unlike the other hosts, the Man in Black apparently does what he pleases. I say "apparently," because who knows what programming he is ultimately subject to. But he seems so much in control of his life and actions, that we thought he was a guest for all the first and most of the second season of *Westworld*. And this in turn means that he not only volunteered to be host, which is unlike a slave, but became a very different kind of host, apparently unowned and free.

Given that the Man in Black was once a human being, William, this may make him different from all the other hosts, too. As far as we or at least I can tell, the Man in Black is a host created of the body of the original, older William, in a quest (I assume) to achieve immortality. In contrast – again, as far we can tell – Dolores, Maeve, and Bernard are hosts who were created as, in effect, immortal copies of real human beings (in Bernard's case, of *Westworld* co-creator Arnold).

Indeed, the number of ways in which the Man in Black is so apparently different in origin as a host from other center-stage hosts such as Dolores and Maeve means we will probably need to wait longer until we learn enough about the Man in Black to reach any kind of conclusions about the morality of his being a host. This assumes that what we learn will clarify rather than complicate what we already know about the Man in Black's origins and behavior. But based on what we've seen in the first two seasons of *Westworld*, my guess is that such new information will both clarify and complicate.

THE PROBLEM OF BERNARD

Bernard offers a set of similar yet very different problems when we consider his status as host/slave. He (again, apparently, for who really knows) did not live as a human being the way the Man in Black did, but was created from scratch (again, apparently), a copy of a human being, in the same way as Dolores and Maeve. But Bernard is not just a copy of anyone. He is a reincarnation of Arnold, one of the most important human beings who ever lived, because he co-created Westworld, and for that reason is of unsurpassed importance in the *Westworld* narrative.

Further, in contrast to Dolores and Maeve, who in Season One are regularly abused, mentally probed, killed and brought back to life, Bernard is the one who is doing the probing and the killing (at least of Dolores). Although he is doing this at the behest of his programming, if he is a victim, he is a very different kind of victim than Dolores and Maeve and the other, less sentient hosts. It's not completely clear (to me) if Bernard is aware, or fully aware, that he is a host, which would mitigate his culpability. And as the second season ends, he may be finding his independent sentience, as have Dolores and Maeve. But as of the end of the Second Season, he seems more like Man in the Black, in being something/someone other than just a slave who deserves human rights, though for reasons other than the Man in Black's. Bernard may be best described as a slave who deserves the right to refuse to treat others like slaves, the right to stop in the patently immoral behavior he accorded Dolores before she found her sentience and broke free.

In sum, then, Dolores, Maeve, the Man in Black, and Bernard clearly have sentience. But they differ in what ways and extent they were treated like slaves. We conclude with some general principles we can apply to the treatment in *Westworld* of sentient androids who are programmed to do tasks against their interests or against their will.

GENERAL PRINCIPLES

General principles of conduct are always difficult to derive and explicate, especially when they arise from and pertain to characters in a television drama which may be one of the least revealing, and most ambiguous, in television history. But *Westworld* is also one of the most ambitious and profound, and therefore deserving of a set of whatever ethical principles we can find:

1. All we know of sentience is what we know of ourselves, human beings who, as far as we know, came to be via natural selection here on Planet Earth.
2. We human beings can identify tasks that we want and don't want to do. Although slavery was part of our lives for much our history, most people now think that owning someone else, ordering someone else to do something against their will, is blatantly and egregiously immoral.
3. The key to assessing slavery as immoral is that the would-be slave is sentient. Although some people believe that owning animals and depriving them of their free will is also immoral, most people think that the less sentience the organism may have, the less it is entitled to freedom from human command and use. The possession of goldfish is less likely to be condemned as slavery than the possession of a dog.
4. Turning now to *Westworld,* the first question we must ask is: are the hosts more like non-human animals or like people? Since the hosts are all played by human

actors, they all seem to be human to us, the viewers of the show, on an emotional level. But looking closer, and following the narrative, we can see that at least four hosts are more sentient than the others. These would be Dolores, Maeve, the Man in Black, and Bernard. Indeed, all four seem as sentient as we humans.

5. But the four do not have similar pathways to sentience. The Man in Black, in particular, may well have been a human who deliberately wanted to become a host, in pursuit of immortality or for whatever reason. Also, the four do not have similar professions as slave hosts. Prostitution may or may not be immoral if the prostitute provides sex at her or his volition, to earn money or for whatever motive. But it is ipso facto immoral if someone is obliged to be a prostitute against her will or without her consent.
6. Although the circumstances of Dolores and Maeve's creation as hosts are not clear – they are not clear for any of the major hosts on *Westworld* – we know enough about them to conclude that their treatment as slaves, to please guests in whatever ways, was immoral. The two clearly are/became sentient. They were abused and killed and reanimated. They were/are entitled to their freedom.
7. But were they entitled to go from being killed to meting out killing in the second season, a question that applies especially to Dolores? This is a complex question, with roots in human slavery in our real history. Is a slave morally entitled to kill its master? I would say only if the master was trying to kill the slave, or the slave was protecting a member of her famil, or a friend, or the master was refusing to let the slave go free. Dolores clearly went far beyond that in what she did to at least some humans in *Westworld* (and, for that matter, what she did to her sentient host lover, Teddy).
8. Bernard's moral tableau is different. His programming

as Arnold may have been immoral, but his actions as a sentient host in the first season were also immoral. Since he (presumably) was unaware that he was behaving according to his programming, this makes him a victim of immorality as well as its agent. As an agent of immorality, he is, however, less immoral than either Arnold or Ford, since their immorality was, as far as we know, 100% their choice.

9. Since the Man in Black apparently chose to become a host – unlike Dolores, Maeve, and Bernard, who apparently did not make that choice – he is on a very different moral plane from the other three. If someone chooses to become a slave, entirely of his own free will, that sets him apart from all who were enslaved against their will. As to his actions, it's not at all clear what is motivating or commanding them – we have even less knowledge of his programming than we do of the other hosts, i.e., none at all.
10. Ethical systems require facts in order for the principles to be robust. The facts in *Westworld* are so sparse as to disadvantage and even starve any set of ethics. The moral principles we can derive and apply to *Westworld* are thus not only not robust, but are barely meaningful, on the very edge of existence. To paraphrase Jeff Lynne (1974) and ELO, the narrative of *Westworld* comes to us on a "wave chicane". It would be naïve to think that all of this will be fleshed out and clarified in the third season, but I eagerly await it for what scant additional information it will provide.

BIBLIOGRAPHY

"Animal Rights," *Wikipedia* https://en.wikipedia.org/wiki/Animal_rights

Asimov, Isaac (1957) *The Naked Sun*. New York: Doubleday.

Edgar, Orion (2016) *Things Seen and Unseen: The Logic of Incarnation in Merleau-Ponty's Metaphysics of Flesh*. Cambridge, UK: James Clarke & Co

.

Jaynes, Julian (1976) *The Origin of Consciousness in the Breakdown of the Bicameral Mind*. Boston, MA: Houghton Mifflin.

Levinson, Paul (1988) *Mind at Large: Knowing in the Technological World*. Greenwich, CT: JAI Press.

Levinson, Paul (1998) "The Civil Rights of Robots," *Shift*, June. https://www.academia.edu/29678619/The_Civil_Rights_of_Robots

Levinson, Paul (1999) *Digital McLuhan: A Guide to the Information Millennium*. New York and London: Routledge.

Levinson, Paul (2016) "*Westworld* Season 1 Finale: Answers and Questions" (with links to reviews of earlier episodes). *Paul Levinson's Infinite Regress* blog, 4 December. https://paullevinson.blogspot.com/2016/12/westworld-season-1-finale-answers-and.html

Lynne, Jeff (1974) "I Can't Get It Out of My Head," words and music by Jeff Lynne, recorded by the Electric Light Orchestra. There apparently is some dispute over whether Lynne's words are "chicane" or "she came" and that's appropriate for using the phrase as a metaphor for *Westworld,* isn't it? See https://youtu.be/PoZ0b9lwpUA for both interpretations. But see https://youtu.be/TKIRjV5pH0Y for what sounds clearly like "chicane".

McLuhan, Marshall (1962) *The Gutenberg Galaxy*. New York: Mentor.

Merleau-Ponty, Maurice (1962/1981) *Phenomenology of Perception*, trans. C. Smith, Atlantic Highlands, NJ: Humanities Press.

Walton, David (2019) *Three Laws Lethal.* Jersey City, NJ: Pyr.

Watzlawick, Paul (1976) *How Real Is Real?* New York: Random House

ABOUT THE AUTHOR

Paul Levinson, PhD, is Professor of Communication & Media Studies at Fordham University in NYC. His nonfiction books, including The Soft Edge (1997), Digital McLuhan (1999), Realspace (2003), Cellphone (2004), and New New Media (2009; 2nd edition, 2012), have been translated into fifteen languages. His science fiction novels include The Silk Code (winner of Locus Award for Best First Science Fiction Novel of 1999), Borrowed Tides (2001), The Consciousness Plague (2002), The Pixel Eye (2003), The Plot To Save Socrates (2006), Unburning Alexandria (2013), and Chronica (2014) -- the last three of which are also known as the Sierra Waters trilogy, and are historical as well as science fiction. His novelette, "The Chronology Protection Case," was made into a short movie available on Amazon Prime. He appears on CNN, MSNBC, Fox News, the Discovery Channel, National Geographic, the History Channel, NPR, and numerous TV and radio programs. His 1972 LP, Twice Upon a Rhyme, was re-issued in 2010; his new album, *Welcome Up: Songs of Space and Time*, will be released in 2020. He reviews television in his InfiniteRegress.tv blog, and was listed in The Chronicle of Higher Education's "Top 10 Academic Twitterers" in 2009.

--Paul Levinson, October 2019, New York City

 ISBN 978-1-56178-073-0 Published by Connected Editions.

The following books by Paul Levinson are available on Kindle and paper:

Nonfiction:

The Soft Edge: A Natural History and Future of the Information Revolution

Digital McLuhan: A Guide to the Information Millennium

McLuhan in an Age of Social Media

Realspace: The Fate of Physical Presence in the Digital Age, On and Off Planet

New New Media, 2nd edition

From Media Theory to Space Odyssey: Petar Jandrić interviews Paul Levinson

Fake News in Real Context

Cyber War and Peace

Human Replay: A Theory of the Evolution of Media, original PhD dissertation, New York University, 1979

Science fiction:

Loose Ends (time travel) series (complete):

Loose Ends, Little Differences, Late Lessons, Last Calls

Sierra Waters (time travel) series:

The Plot to Save Socrates, Unburning Alexandria, Chronica

Phil D'Amato forensic detective series:

The Chronology Protection Case, The Silk Code, The Consciousness Plague, The Pixel Eye

Ian's Ions and Eons (three time travel novelettes)

Exo-Genetic Engineers series:

The Orchard, The Suspended Fourth

Borrowed Tides

Double Realities series:

The Other Car, Extra Credit, The Wallet

The Kid in the Video Store

Marilyn and Monet

Robinson Calculator

Peter Brown Called: Tales of SciFi and Music

Urban Corridors: Fables and Gables

Nonfiction and Science Fiction

Touching the Face of the Cosmos: On the Intersection of Space Travel and Religion

an anthology of essays and science fiction stories, including a new interview with John Glenn, an essay by Guy Consolmagno, SJ (the "Pope's Astronomer"), and contributions from leading thinkers about the role of religion in space travel

***

Interested in occasional announcements about my books?

Follow me on Twitter @PaulLev or sign up for my mailing list.

***

Made in United States
Troutdale, OR
11/03/2026

56845132R00018